all   the   New

Please bring us some figgy pudding,
Please bring us some figgy pudding,
Please bring us some figgy pudding,
Please bring it right here.
Good tidings to you and to all your kin,
Good tidings for Christmas and all the New Year.

We won't leave until we get some,
We won't leave until we get some,
We won't leave until we get some, Please bring it right here.
Good tidings to you and to all your kin,
Good tidings for Christmas and all the New Year.

Silent Night

Sleep ___ in heav - en - ly peace_____.

Silent Night, holy night!
Shepherds quake at the sight.
Glories stream from heaven afar,
Heavenly hosts sing, "Alleluia!"
Christ, the Savior is born,
Christ, the Savior is born.

Silent night, holy night!
Son of God, love's pure light!
Radiant beams from Thy holy face
With the dawn of redeeming grace.
Jesus, Lord at Thy birth,
Jesus Lord at Thy birth.

Jingle Bells

Oh, what fun it is to ride a one - horse o - pen sleigh! Oh,

Jin - gle bells! Jin - gle bells! Jin - gle all the way!

Oh, what fun it is to ride a one - horse o - pen sleigh!

A day or two ago, I thought I'd take a ride,
And soon Miss Fannie Bright was seated by my side;
The horse was lean and lank, misfortune seemed his lot,
He got into a drifted bank, and we, we got upsot.

Oh, Jingle bells! Jingle bells! Jingle all the way!
Oh what fun it is to ride a one-horse open sleigh!
Oh, Jingle bells! Jingle bells! Jingle all the way!
Oh what fun it is to ride a one-horse open sleigh!

Now the ground is white, go it while you're young,
Take the girls tonight, and sing this sleighing song;
Just get a bobtailed nag, two-forty for his speed,
Then hitch him to an open sleigh, and crack! You'll take the lead

Oh, Jingle bells! Jingle bells! Jingle all the way!
Oh what fun it is to ride a one-horse open sleigh!
Oh, Jingle bells! Jingle bells! Jingle all the way!
Oh what fun it is to ride a one-horse open sleigh!

Deck the Halls

See the blazing Yule before us,
Fa la la la la la la la la.
Strike the harp and join the chorus,
Fa la la la la la la la la.
Follow me in merry measure,
Fa la la la la la la la la.
While I tell of Yuletide treasure,
Fa la la la la la la la la.

Fast away the old year passes,
Fa la la la la la la la la.
Hail the new, ye lads and lasses,
Fa la la la la la la la la.
Sing we joyous all together,
Fa la la la la la la la la.
Heedless of the wind and weather,
Fa la la la la la la la la.

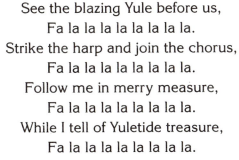

O Little Town of Bethlehem

met in thee to - night.

For Christ is born of Mary and gathered all above
While mortals sleep, the angels keep their watch of wondering love.
O morning stars, together proclaim the holy birth,
And praises sing to God the King and peace to men on earth.

How silently, how silently the wondrous gift is given,
So God imparts to human hearts the blessings of his heaven,
No ear may hear His coming, but in this world of sin,
Where meek souls will receive Him, still the dear Christ enters in.

O holy Child of Bethlehem descend to us, we pray,
Cast out our sin and enter in, be born in us today.
We hear the Christmas angels the great glad tidings tell,
O come to us with us our Lord Emmanuel!

It Came Upon a Midnight Clear

It came up - on a mid-night clear, That

glo - rious song of old, From an - gels bend - ing

near the earth to touch their harps of gold. "Peace

on the earth good will to men, From heav'ns all gra - cious

King." The world in sol - emn still - ness lay to

hear the an - gels sing.

They looked up and saw a star
Shining in the east beyond them far,
And to the earth it gave great light,
And so it continued both day and night;
Noel, Noel, Noel, Noel,
Born is the King of Israel!

And by the light of that same star
Three Wise Men came from a country afar,
To seek for a king was their intent,
And to follow the star wher'er it went;
Noel, Noel, Noel, Noel,
Born is the King of Israel!

The First Noel

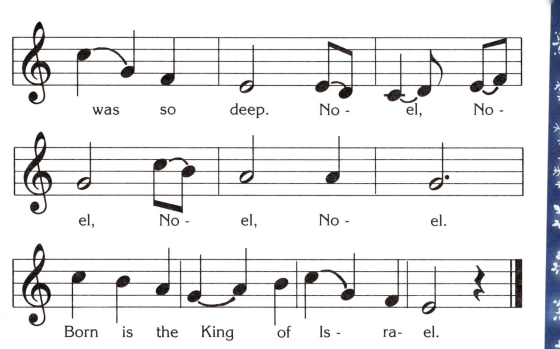

was so deep. No - el, No -

el, No - el, No - el.

Born is the King of Is - ra- el.

They looked up and saw a star
Shining in the east beyond them far,
And to the earth it gave great light,
And so it continued both day and night;
Noel, Noel, Noel, Noel,
Born is the King of Israel!

And by the light of that same star
Three Wise Men came from a country afar,
To seek for a king was their intent,
And to follow the star wher'er it went;
Noel, Noel, Noel, Noel,
Born is the King of Israel!

Hark The Herald Angels Sing

Hark! The her - ald an - gels sing, Glo - ry to the

new - born King; Peace on earth and mer - cy mild,

God and sin - ners re - con - ciled! Joy - ful all ye

na - tions rise, Join the tri - umph of the skies;

With th'an - gel - ic hosts pro - claim, Christ is born in